Remarkable Robots

By Anastasia Suen

CELEBRATION PRESS

Pearson Learning Group

Contents

What Is a Robot?

If someone asked you to draw a robot, you might sketch a shiny, silver figure that looked vaguely like a human. That's what the robots in old movies and TV shows looked like. These humanlike machines, called **androids**, walked stiffly and spoke in a strange way. They did whatever humans told them to do.

In movies today, the robots often look like humans. They have skin and hair, and they speak just like anyone else. In fact, sometimes they're smarter than humans!

Neither type of movie robot exactly fits the real definition of a robot. A robot is simply a machine that moves. However, movie robots tell us something true about human nature: People have long dreamed about machines that would look like humans and do their work.

This android, "C-3PO," was first introduced in the 1977 movie *Star Wars*.

3

The word *robot* actually came from a play about machines that looked like humans. In 1921, Karel Capek wrote a play called *R.U.R.* (Rossum's Universal Robots). In this play, shiny, silver machines that looked like humans did the work that humans themselves didn't want to do.

Even the word *robot* has to do with work. Capek was Czech, and the Czech word *robota* means "forced labor." The robots in his play did the work they were forced to do. After the play was performed, people began to use the word *robot*.

The idea of machines that look like humans wasn't new. For as long as machines have existed, people have been figuring out ways to make them look like humans or animals. This has been true since the days of ancient Greece. Homer wrote about golden machines that looked like humans. These machines helped one of the Greek gods do his work. Around 400 B.C., Archytas (ar-KY-tus), a mathematician, made a revolving machine that looked like a dove. Arab, Chinese, and Greek inventors made machines that looked like human or animal figures. Flowing water made the figures move.

Machines that move by themselves are called **automatons**. Early automatons used water for power. Then inventors discovered springs and gears. In the 1500s, an inventor named Hans Bullmann used springs and gears to create the first androids. Some of them played musical instruments.

Machines that moved from place to place came next. In the 1700s, Japanese craftspeople invented androids that moved from place to place. These androids looked like dolls, with clothes and hair. When you placed a teacup in the doll's hand, it rolled across the room to serve tea. When your guest picked up the teacup, the doll stopped. When the guest put the teacup back in the android's hands, it turned around and rolled back to you.

What is the difference between these early androids and today's androids? Automatons move because of their gears and springs. To make an early android move a different way, you would need to change the way the springs and gears connect. Today's androids, on the other hand, just need a new software program to tell them to move in a different way. This is because modern robots use computers for their "brains." Robots as we know them today came into existence after computers were invented.

George Devol invented the first programmable robot. Devol joined forces with another American, Joseph Engelberger. They developed the Unimate, which went to work in 1961. Like all early-modern robots, Unimate was just an arm.

Unimate worked for General Motors. At the GM factory, hot metal was poured into a mold to make car

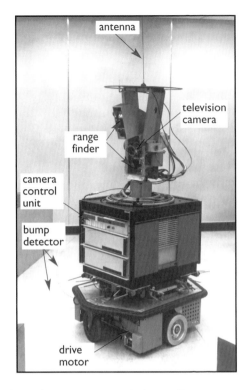

The Shakey robot

antenna

television camera

range finder

camera control unit

bump detector

drive motor

parts. The robot took the hot car parts out of the mold.

Soon, scientists began using robots in space. In 1967, NASA sent *Surveyor 3*, a **probe**, to the Moon. Its TV camera sent pictures to Earth, while its robot arm tested the soil.

The next developments in robots were closely linked to changes in computer technology. An early robot called Shakey gathered information using a radio and TV camera. It sent the information to a large computer. Then Shakey decided where to move next.

When a robot uses a program to think, as Shakey did, we say that it has **artificial intelligence**. Shakey was a big breakthrough. In the decades to follow, people would invent robots that could walk, crawl, fly, and even play soccer!

The History of Robots

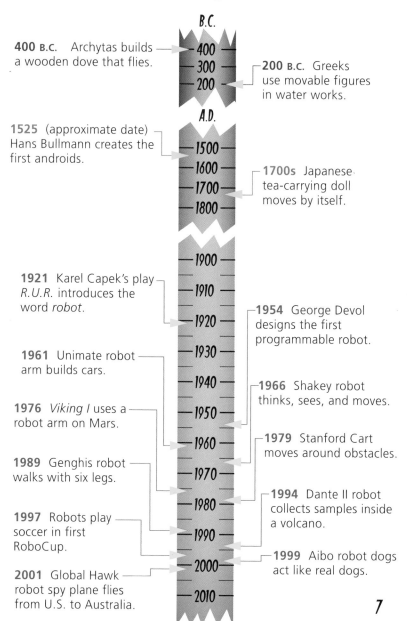

B.C.
- 400
- 300
- 200

400 B.C. Archytas builds a wooden dove that flies.

200 B.C. Greeks use movable figures in water works.

A.D.
- 1500
- 1600
- 1700
- 1800

1525 (approximate date) Hans Bullmann creates the first androids.

1700s Japanese tea-carrying doll moves by itself.

- 1900
- 1910
- 1920
- 1930
- 1940
- 1950
- 1960
- 1970
- 1980
- 1990
- 2000
- 2010

1921 Karel Capek's play *R.U.R.* introduces the word *robot*.

1954 George Devol designs the first programmable robot.

1961 Unimate robot arm builds cars.

1966 Shakey robot thinks, sees, and moves.

1976 *Viking I* uses a robot arm on Mars.

1979 Stanford Cart moves around obstacles.

1989 Genghis robot walks with six legs.

1994 Dante II robot collects samples inside a volcano.

1997 Robots play soccer in first RoboCup.

1999 Aibo robot dogs act like real dogs.

2001 Global Hawk robot spy plane flies from U.S. to Australia.

Factory Robots

The first robots worked in factories. These were called **industrial** robots. They made things.

Industrial robots do work that people cannot do or don't want to do. This includes work that is highly repetitive. Industrial robots also do work that is too dangerous for humans.

The very first robots made cars, and robots still work in car factories today. They cut car parts, inspect them, join them together, and perform other tasks. Other industrial robots load and unload machines. They clean parts, paint them, and do many other jobs. Robot arms do most of this work.

Robots are used in industries everywhere. At the end of 2000, the United Nations reported that about 750,000 industrial robots were being used worldwide.

Industrial Robots in 2000

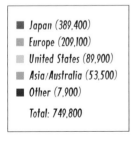

- Japan (389,400)
- Europe (209,100)
- United States (89,900)
- Asia/Australia (53,500)
- Other (7,900)

Total: 749,800

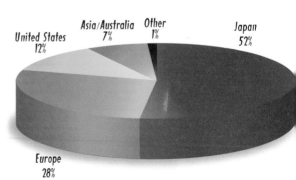

Japan 52%
Europe 28%
United States 12%
Asia/Australia 7%
Other 1%

Robots are still making cars.

Industrial robots don't look like humans, but they are like people in some ways. A robot has joints to connect its body parts. So do you. A robot uses motors to move. You use your muscles. Robots use a power source to give them energy. You eat food for energy. The robot's program tells it what to do. You use your brain to make decisions.

Using your five senses comes naturally to you, but not to a robot. In order for a robot to notice something, it must have a **sensor**. Robots use sensors to "sense" the world around them. One type of sensor helps the robot see. Another helps the robot figure out how close an object is.

Touch sensors help robots to pick up objects without breaking them. "Pick and place" robots work in assembly lines. They see a chocolate on the conveyor belt and pick it up to place it in the candy box.

Robots move by bending their joints, just like you do. Most robot arms can move in six different ways. These six movements are called the *six degrees of freedom*.

Hold your arm out. Then move it back in. This is the first degree of freedom, moving *in and out*.

Point to the ceiling and then the floor. The second degree of freedom is moving *up and down*.

Use your right hand to touch your left shoulder. Now touch your right shoulder. Moving *left and right* is the third degree of freedom.

Next, use your wrist. Make your palm face the ceiling. Then twist it to face the floor. This movement, a *roll*, is the fourth degree of freedom.

Keep your arm steady, and twist your wrist to point your fingers at the floor and then the ceiling. This *pitch* is the fifth degree of freedom.

Now twist your wrist left and right. This *yaw* is the sixth degree of freedom.

The six degrees of freedom are not limited to robots. Pilots use the words *roll*, *pitch*, and *yaw* to describe the movement of planes in the air. The words also describe how a ship moves in the water.

Not all robots have six degrees of freedom. It depends on the task they need to do. Some robots have only four degrees of freedom.

Although a robot arm is like your arm in some ways,

the two are not the same. Your shoulder is attached to a body. A robot arm may be the robot's entire "body."

If a robot has only an arm, then the robot's "brain" can't be located in its head. The computer that thinks for the robot is outside the robot arm.

A robot's "hand" is also different. You have five fingers, but robots with fingers have only two or three. A robot's tools are attached right to its hand.

What kind of "hand" the robot has depends on the kind of work it does. Robots that spray paint on cars have different kinds of hands than robots that put chocolates into boxes.

The Six Degrees of Freedom

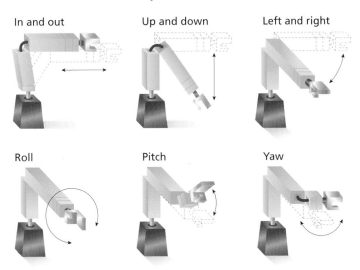

In and out

Up and down

Left and right

Roll

Pitch

Yaw

Explorers

Robot arms stay in one place to do their work, but robots that are bolted to the floor can't explore the world. That's why **roboticists**, the people who make robots, decided to add wheels to their robots.

At first, exploring a room was enough. That's what Shakey did. Soon, the robot makers were ready to move outdoors. The Stanford Cart was an early robot that did just that. It looked like a computer table with bicycle wheels.

Scientists tested the Stanford Cart in the 1970s. The robot's computer "brain" told it how to move, but it was slow. It took 10 to 15 minutes of stops and starts for the robot to roll a meter, just 39.4 inches. After each meter, it stopped again to think about where to move next.

Wheels worked on flat surfaces, but what about bumpy ones? Insects are good at climbing over things, so robot makers decided to imitate insects. They made robots that moved on many legs.

Genghis was one of the first robots with legs. Like an insect, Genghis had six legs. It was only 7 inches tall. Attila and Hannibal were next. Built in the early 1990s, these "insect" robots were designed to explore other planets. The small insectlike robots

Dante II climbed inside a volcano.

could sense objects and then climb over them.

After these small walking robots came large walking robots. The two Dante robots each had eight legs—four on each side. Like a mountain climber, each Dante robot used a cable to lower itself into a volcano. Dante I explored a volcano in Antarctica. Dante II explored the Mount Spurr volcano in Alaska.

Robots move in two ways. Some move with remote control. Humans decide how the robots move. Other robots are **autonomous** and move by themselves. Their programs help them decide what to do next.

Remote-controlled robots work just like remote-controlled toys. When you hold the controls, you are in charge. The robot's program is inside the robot, but starting and stopping the program is up to the human at the controls. If the robot needs to go in a different direction, the human makes the changes, not the robot.

Remote-controlled robots allow humans to "visit" places that humans cannot go by themselves. In 1993, the TROV robot swam under the sea ice in Antarctica while scientists in California operated the controls. The TROV did scientific experiments as it explored the waters 800 feet below the surface.

The people at the controls were watching pictures from the robot's cameras. They also read information from the robot's sensors. Then they used controls to move the TROV.

A remote-control robot can be changed into an autonomous robot that moves by itself. The Nomad **rover** is a good example. In 1997, the Nomad traveled across a desert in Chile. Scientists in California used remote control to make the Nomad do experiments. At the same time, students in Pennsylvania took turns driving the rover. Computer networking made this possible.

In 2000, Nomad was taken to Antarctica to hunt for rocks that had fallen from outer space. This time, scientists changed the robot's program. They gave the robot artificial intelligence. Now, when the robot took in information from its sensors, it didn't send the information back to a human for a decision. It sent the information to another program which decided what to do next.

Did Nomad find any space rocks, or meteorites?

The Nomad rover in Antarctica

It found five, but it had some problems. It had trouble telling some Earth rocks from space rocks.

Testing robots such as Nomad in extreme conditions, helps scientists design robots that can be used in outer space. *Surveyor 1* was the first American robot to land on the Moon. It landed in 1966, three years before humans went there. NASA used remote control to operate *Surveyor 1*. The robot conducted experiments and sent the information back to Earth.

Today, NASA sends autonomous robots into space. Robots sent to work on other planets need to be able to work by themselves. It can take from 4 to 20 minutes for a signal to travel from Earth to Mars. This makes it too hard to drive a robot by remote control.

Autonomous robots in space do have remote controls for some jobs. If a robot finds something interesting, a scientist can use remote controls to send the robot back for another look.

Humans haven't been to Mars yet, but robots have. NASA sent twin landers to Mars. *Viking 1* was launched on August 20, 1975. It didn't land on Mars until July 20, 1976! The *Viking 1* and *Viking 2* landers used their robotic arms to do experiments on Mars.

In 1997, *Mars Pathfinder* landed on Mars. Inside this lander was a rover robot called *Sojourner*. The rover drove around Mars and did experiments. It sent 3-D pictures back to the lander, which sent the pictures to Earth. Scientists used 3-D glasses (with one red lens and one blue lens) to see the pictures.

In 2003, NASA is sending twin rover robots to Mars. The rovers will land in two different places on Mars to do experiments. Once they reach the planet, the rover robots will be independent of their landers. They will send pictures back to Earth by themselves.

The Mars exploration rovers will land on Mars in 2004.

Spies

Did you ever want to be a spy? You could watch people without anyone knowing you were there. You could go behind enemy lines to find out important secrets!

Robots are sometimes used as spies. They are very valuable in this work, because robot spies can go where humans cannot.

During a war, it's important to know what the other side is doing. In the Vietnam War, the U.S. military sent aircraft behind enemy lines to take photographs. They used the information to see where the enemy's troops were located.

This was dangerous work. Many planes were shot down. The photographs were important, but the human cost was too high. The military decided to use planes without pilots.

The military had been using unmanned planes for years. In the 1930s, small planes were flown by remote control and used for target practice. These target planes were known as target **drones**. A drone, or male bee, doesn't sting. Target planes didn't "sting," either. They didn't fight back. The drones were launched into the air from aircraft carriers, using catapults.

During the Vietnam War, Firebee target drones were turned into robot spy planes. Cameras were added, and the drones flew out over enemy lines to take pictures. These robot spies had a code name. They were called Lightning Bugs.

Each robot was programmed to fly to a specific site and take photographs. The robots couldn't take off by themselves. They were attached to a large cargo plane that dropped them off in midair. A cargo plane could hold four drones at once.

After the robot spies took their photographs, they were programmed to come back and land. Drones didn't land at an airport. A parachute opened and the robot dropped from the sky.

Some parachute landings turned into crash landings. Robots that crashed couldn't be used again, so a new plan was developed. A large helicopter flew out with a net to catch the robot before it hit the ground. Then the robot could be used again. From 1964 on, robot spies flew thousands of missions during the Vietnam War.

During the Gulf War in 1990 and 1991, U.S. aircraft carriers had their own robot planes. Their spy robot was called the Pioneer. A Pioneer robot plane was launched from a ship to fly behind enemy lines. The robot plane was flown by remote control from the aircraft carrier.

The Pioneer robot plane was launched from a ship.

During the Gulf War, some Iraqi soldiers surrendered to a Pioneer robot! When they saw the robot flying over their heads, the soldiers knew that they had been discovered. They waved white flags at the robot to show that they did not want to fight. Waving a white flag on a battlefield is the universal sign of surrender.

When the Pioneer robots finished their work, they flew back to the ship to land. A ship moves up and down and side to side because of the waves. The safest way to land a robot on a ship is to fly it into a huge net. After the net stops the robot, it is lowered to the deck and put away.

A robot that flies is also called a UAV. UAV stands for **Unmanned Aerial Vehicle**. That means it flies in the air by itself and carries things from place to place.

Pioneer isn't the only robot spy today. The military is developing new robot spies, both larger and smaller than the Pioneer.

The Global Hawk UAV is a large robot spy. It takes pictures of large land areas. In fact, it can look at an area the size of Illinois! The wings are 116 feet across.

The Global Hawk is autonomous. It rolls down the runway, takes off, and flies for more than a day without anyone telling it what to do. When the mission is complete, it comes back to base and lands by itself.

In 2001, Global Hawk set a record when it flew from the United States to Australia. No robot had ever flown 7,500 miles without stopping.

Smaller robot spies, like the Dragon Runner, can fit in a backpack. The Marines are testing this robot for combat. It looks like a toy car without windows. Instead of headlights, it has a camera.

The Dragon Runner weighs only about 16 pounds. To use it, a Marine grabs its flexible handle and throws the robot through a window, up stairs, or over a wall. However it lands, the sensors in the robot know which way is up. Then the Dragon Runner can be driven by remote control. Dragon Runner can also sense motion. If a person is nearby, the remote control vibrates like a pager.

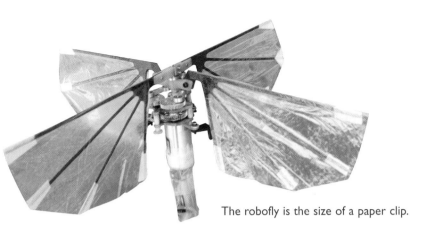
The robofly is the size of a paper clip.

Have you ever wanted to hear something that was being said in another room? Did you wish you could "be a fly on the wall" of the room? There's a tiny flying robot that can gather information, and it looks like a fly! Scientists call it a robofly.

If you've ever chased a fly, you know that flies can take off and land upside down, right side up, or even sideways. They use several different wing motions. Houseflies have only two wings, but to make a robot fly, scientists had to use four wings.

The robofly can flap its wings more than 150 times a second. It doesn't need robot fuel—or even food like a real fly. The robofly uses the sun for power.

Scientists are still testing roboflies. They hope to send out a large group of the robots on one mission. If some are damaged, the others can still send back information. Roboflies can be used in cities, inside buildings, even on other planets.

A Helping Hand

A doctor attends school for many years to become a medical expert. A robot can't replace a doctor, but some robots work with doctors during surgery.

One procedure that uses robots is heart surgery. A heart operation is quite a shock to the body. In order to reach the heart, the surgeon must cut through skin, bones, muscles, veins, and nerves. During surgery, clamps hold the ribs apart. The heart itself must be stopped. The patient is kept alive on a heart-lung machine while the doctor operates.

If the doctor uses a robot to help with the surgery, he or she only needs to cut tiny openings in the body. The robot's arms are as thin as a pencil. They can work in an opening that is just 1 centimeter wide, or less than half an inch.

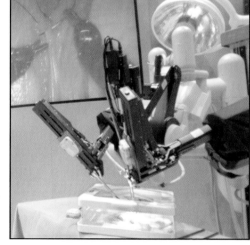

This remote-controlled surgical robot has three arms.

A tiny camera goes into one opening. This camera allows the doctor to see inside the body. The camera also makes everything look larger.

The other openings are for the robot's skinny arms. The robot's arms have tools instead of hands. How does it know how to use those tools? For some surgery, doctors direct the robot by using remote control. The doctor moves joysticks, just as you would do to control an electronic game. The robot copies the doctor's movements inside the body.

For other kinds of surgery, robots use **voice recognition**. The robot's computer memory is programmed to recognize the doctor's voice commands. When the doctor speaks clearly into a headset, the robot does what the doctor says. For instance, if the doctor wants more light, the command might be, "Lights up."

A surgery robot may use both voice recognition and joystick controls. The camera's movement responds to the doctor's voice commands, and the surgery arms move when the doctor moves the joysticks. Robot arms don't shake as human arms do when they stay in the same place for a long time.

Patients who have the regular heart surgery take months to recover. Patients who have robot surgery recover much more quickly. There is less damage to tissues, muscles, nerves, and bones.

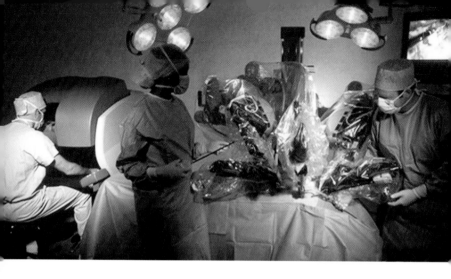

A robotic telesurgical work station

Surgery robots are also used in **telemedicine**. *Tele* is Greek for "at a distance." Telemedicine happens when a doctor helps a patient who is far away. It has been used since the early days of radio in the 1920s. Doctors on shore would help patients on ships by talking on ship-to-shore radios. The crew on the ship helped the patient by following the doctor's instructions.

With telemedicine, robots can help people across town, across the country, or across the ocean. In 2001, a doctor in the United States used a robot to remove the gallbladder of a woman in France. The doctor in the United States controlled three robot arms. The robot and the patient were in an operating room with doctors in France.

Robots also lend a helping hand to people who are disabled. Some people need to hold on to something when they walk. A robot walker can help. When a person holds the handlebars of a robot walker, it rolls forward. The robot sensors make sure that the walker does not bump into things. Because the robot also tells what is coming, it can be used as a guide for blind people.

Many wheelchairs already use a joystick to help the person steer. For people who cannot use their arms very well, steering with a joystick is difficult. A wheelchair robot can figure out where the person is trying to go when he or she moves the joystick. It moves in that direction and at the same time, it avoids bumping into things.

Some wheelchairs have a robot arm. The arm can pick things up, turn on a light, or press an elevator button. It helps people to help themselves.

A robot walker helps to guide a blind person.

Robots that can help around the house have been a dream for centuries. Imagine how much easier your life would be if a personal robot could do some of your work!

This dream is not a reality yet. Most robots do the same thing over and over. At home, you do many different kinds of tasks. Washing the car is not the same as washing the dishes or washing clothes. You wash all three with soap and water, but the movements and the work places are not the same.

Training a robot to do random movements has been harder than robot makers expected. Moving in many different directions is not as easy as just going back and forth over and over again.

At the same time, machines for the home have been invented to do the work that robots might do. Most people don't chop wood for a fire to cook dinner, for example. They use an oven.

Today's personal robots only do things that other machines already do. There are vacuum-cleaning robots and lawn-mowing robots. These robots still need people to help them work. For instance, you have to charge the lawn-mowing robot's batteries for 24 hours before you use it. You also have to clean up the yard! When it comes to housework, robots are a long way from replacing people.

Entertainers

In 1939, two robots appeared at the World's Fair in New York. Elektro was a giant robot that looked like a man. Sparko, a robot dog, was his pet.

Elektro and Sparko appeared on stage. Elektro could walk and talk and dance. His dog, Sparko, walked and sat and barked. These early robots didn't build or explore or spy or help anyone. Their only job was to entertain people at the fair.

Today, you can find entertainment robots at amusement parks and museums. At an amusement park, robots play the parts of different characters. Some look like people, while others are dressed up as creatures. These robots may welcome you to an exhibit, give a famous speech, sing and dance, or try to scare

Elektro and his dog, Sparko, appear at the 1939 World's Fair.

you! When the next group of visitors comes in, they do it all over again.

At a museum, you may find dinosaur robots. Unlike the skeletons in the other rooms, these dinosaurs can move! It is quite a change to go from a quiet room filled with bones to a room of moving, roaring robots.

Toy robots also entertain. After the play *R.U.R.* was produced in 1921, toy makers began to make toy robots. Early toy robots were made of tin, just like other toys of the day. To make the toy move, you turned the key to wind it up. Wind-up robots usually marched forward, moving their legs one step at a time. Some robots swung their arms as they marched. After the toy wound down, it stopped. Today, people collect toy tin robots as a hobby.

When mechanical toys began to use batteries, toy robots used them, too. When you flipped the switch, the robot moved. Some battery-powered robots also had remote controls, so you could turn the robot on and off from far away.

As computers changed, so did toy robots. Today's toy robots have computers inside, just like real robots. You can buy a robot kit and make your own robot with a tiny computer inside. Some robot kits allow you to use your personal computer to write a new program for your robot.

This robot cat purrs
when you stroke its fur.

Most of the early robot toys looked like Elektro. They were androids, machines that looked and moved like humans.

Today there are also toy robots that look like animals. One robot dog is called Aibo. This name comes from the words **A**rtificial **I**ntelligence ro**BO**t. This robot dog acts like a real dog. You can teach it to follow your commands. One program for Aibo allows you to help the dog grow up. At first, it acts like a puppy. As you train it, it begins to act like an older dog.

There is also a robot cat. Like real cats, they don't follow directions! These robot cats stretch their legs and move their ears, but they don't walk. The robot cat makes 48 different cat sounds.

Robots also entertain people by playing games. In some games, two robots compete. There are even sumo robots that try to push each other out of a ring, just as sumo wrestlers do.

Several TV shows have robots fighting against each other to see which is the strongest. In order to win, one robot must destroy the other robot. Some people don't like the violence on these shows. Others compare it to boxing, in which two people fight until one is declared the winner.

In other games, the two robots don't touch each other. Instead, one robot tries to score more points than the other. Both robots are given the same task. The one that scores the most points wins the contest.

Middle school, high school, and college students compete in robot contests. They work in teams to build their own robots. In a match, these teams find out how well their robots really work. The robot team waits in the pit area, just as racing teams do at car races. The pit area is where the teams keep their tools and make last-minute repairs to their robot.

Each contest has different rules. These rules tell the robot makers how big to make their robot, what materials to use, and what the robot must do to win. Robots may race like cars, swim, or climb ropes. In the firefighting robot contest, the robot must enter a house and put out a fire.

Soccer-playing robots compete in RoboCup.

Robots also play together as a team. The first Robot World Cup Soccer Games were played in Japan in 1997. Teams from around the world competed. The robots played in leagues based on their size and type.

The Robot World Cup Soccer Games are also called the RoboCup. As robots have changed over the years, robot soccer has added new leagues. In 2002, robot soccer started a league of android robots. By the year 2050, RoboCup wants to have a team of androids that can play soccer against a human team.

Robots have come a long way since 400 B.C., when Archytas made a wooden dove fly. Many of the things people dreamed that robots would do one day have come true. Today, robots work all around the world and even on other planets! Someday there may even be a robot that can do all of your chores. Who knows what robots will do next!

Glossary

androids	robots that look like humans
artificial intelligence	thinking that is done by a machine
automatons	machines that operate by themselves
autonomous	not controlled by others
drones	planes without pilots that fly by remote control
industrial	used in a business that makes products to sell
probe	an unmanned spacecraft used for sending back information from space
roboticist	a person who designs and builds robots
rover	a vehicle used to explore the surface of a planet
sensor	a device that receives information and responds to it
telemedicine	healthcare that uses technology to communicate from a distance
Unmanned Aerial Vehicle	a plane that flies without a pilot, also known as a UAV
voice recognition	the identification of human speech by a machine